Not Strictly by the Numbers

12/16/02

Best wishes,

Bob Knaaff

Not Strictly by the Numbers

Harry Blair
Bob Knauff

Carolina™ Mathematics
Carolina Biological Supply Company
Burlington, North Carolina

Not Strictly by the Numbers

ISBN: 0-89278-150-5
Library of Congress Card Number 90-84844

Carolina™ Mathematics
Carolina Biological Supply Company
2700 York Road
Burlington, NC 27215
USA

1-800-334-5551

Not Strictly by the Numbers

Guaranteed to Be at Least $16^2/_3$% Funny

This book of cartoons is the offspring of a set of twelve cartoons that was published in early 1990, intended for use in mathematics classrooms. We were pleased to find that the cartoons were well received not just by mathematics educators and students but, in fact, by almost everyone who saw them, even by those who profess an aversion to mathematics. That encouragement was all we needed to embark upon the development of this larger collection of cartoons, which attempts to poke fun through a slightly out-of-kilter examination of the somewhat uneasy relationship between humankind and mathematics.

There was one notable exception to the positive reception of the first dozen cartoons—one set was returned with a note that stated: "Only two of the cartoons were funny enough to keep, so I am returning them all." Our colleague, Elizabeth Brazell, comforted us by observing, "Two out of twelve isn't bad. At least they're $16^2/_3$% funny." Ever since, $16^2/_3$% funny has been the standard to which we hold similar products.

Not Strictly by the Numbers begins with eleven of the original cartoons (we deleted one of the funny ones) complemented by a larger set of new ones that, we believe, also meets or exceeds the standard of $16^2/_3$% funny. We submit them for your review and trust that they will engender a measure of enjoyment.

Harry Blair
Bob Knauff

"I DUNNO, JAKE... 38-LITER HAT
JUST DON'T HAVE QUITE THE SAME RING."

"GRADES HAVE GONE UP EVER SINCE WE CHANGED
THE COVERS ON THE MATH BOOKS!"

" I WANTED A <u>NUMBER LINE</u>, NOT A NUMBED LION !! "

"BACK UP A LITTLE MORE. THIS SAYS I'M FOCUSED ON <u>INFINITY</u>."

"YOU EXPECT ME TO BELIEVE THAT'S WHY
YOU DON'T HAVE YOUR MATH HOMEWORK DONE ?!!"

UNEARTHING FOSSILS OF PREHISTORIC NUMBERS

11

"HE COPIED THE PROBLEM WRONG."

" 5¾ CLAMS? HMMM... HOW MUCH THAT IN TIGER TEETH? "

IT WAS A LITTLE KNOWN BUT SOMETIMES USEFUL FACT
THAT DRACULA HAD A SEVERE CASE OF MATH ANXIETY.

DIPLOMATIC SCALE COMPANY

"IT'S ONLY A FLYSPOT,
BUT IT BALANCED THE BUDGET."

"NOBODY DOES FORMAL PROOFS LIKE DR. WINSLOW."

BLAIR

"WHAT'S THE MATTER, CAN'T YOU APES COUNT?"

"THIS TRIBE SEEMS TO HAVE
REMARKABLE BUSINESS SENSE!"

" YOU'VE GOT A <u>LIFETIME</u>, NO-CUT CONTRACT...
NOW GET OUT THERE AND FIGHT ! "

" I'VE INVENTED A NUMBER SYSTEM BASED ON TEN.
IT'S INTERESTING, BUT I'M NOT SURE THERE ARE ANY
PRACTICAL APPLICATIONS . "

"WELL, IT'S 3 B.C....TIME IS RUNNING OUT FOR US TO ATTACH A MEANING TO THIS 'B.C.' THING."

"SAY...WEREN'T YOU MY EIGHTH GRADE
ALGEBRA TEACHER?"

WMTH

BLAIR

"AND NOW THE BASEBALL SCORES...
4 TO 3, 6 TO 2, AND 8 TO 1!"

EXTINCT SPECIES

DODO

SABER-TOOTHED TIGER

APATOSAURUS

SLIDE RULE
REPAIRMAN

BLAIR

" IT'S EASY TO REMEMBER THE COMBINATION.
MICHAEL JORDAN (23) RIGHT... JOE MONTANA (16) LEFT...
JOSE CANSECO (33) RIGHT. "

"PROFESSOR McCABE, I'M AFRAID YOU'VE GOT AN
ADVANCED CASE OF COROLLARY THROMBOSIS."

"WE'LL SHOVE OFF AS SOON AS
I ISOLATE A PAIR OF AMOEBAS."

" GO!"

" I'D CALL THIS A DESIGN FLAW.
THERE'S A TENTHS PLACE ON THE FLOOR SELECTOR. "

"THAT'S NOT WHAT I HAD IN MIND WHEN I ASKED YOLI TO BRING A RULER TO CLASS."

38

39

ARCHIMEDES COMMERCIALIZES HIS PRINCIPLE.

"OUR BALL, OR YOURS?"

"DOUSE THE FIRE, CHATTERING BULL ...
IT'S ANOTHER WRONG NUMBER."

MATH DEPARTMENT LAVATORIES

"WE ASKED YOU A QUESTION, LEFTY ...
ONE MORE TIME NOW ... WHAT'S THE
SQUARE ROOT OF 268,324 ?!! "

"YOUR LAWS OF MOTION WERE GREAT, YOUR
DEVELOPMENT OF CALCULUS WAS INCREDIBLE,
BUT I REALLY LIKED THE COOKIES
WITH THE FIG FILLING ! "

10× MICROSCOPE

"AN INFINITE SERIES?? BUT I HAVE
TO BE HOME BY FIVE O'CLOCK."

"HOW MANY TIMES DO I HAVE TO TELL YOU
NOT TO MIX UNITS OF MEASURE?"

NEWTON HAVING ONE OF THOSE DAYS WHEN
THE IDEAS JUST DON'T COME

"FEE PHI RHO FUM ALPHA BETA GAMMA ..."

AFTER LEAVING THEIR STATIONS AT NOON AND
TRAVELING TOWARD ONE ANOTHER AT 60 MPH
AND 80 MPH, RESPECTIVELY, THE TRAINS
MEET AT 3:08 PM.

" TEN-NESS, ANYONE ? "

"NO, CAMERON, ONE PLUS ONE
DOES NOT EQUAL 'SNAKE EYES'!"

"LOOKS LIKE THE FIFTIES WERE MORE CONSERVATIVE THAN WE THOUGHT."

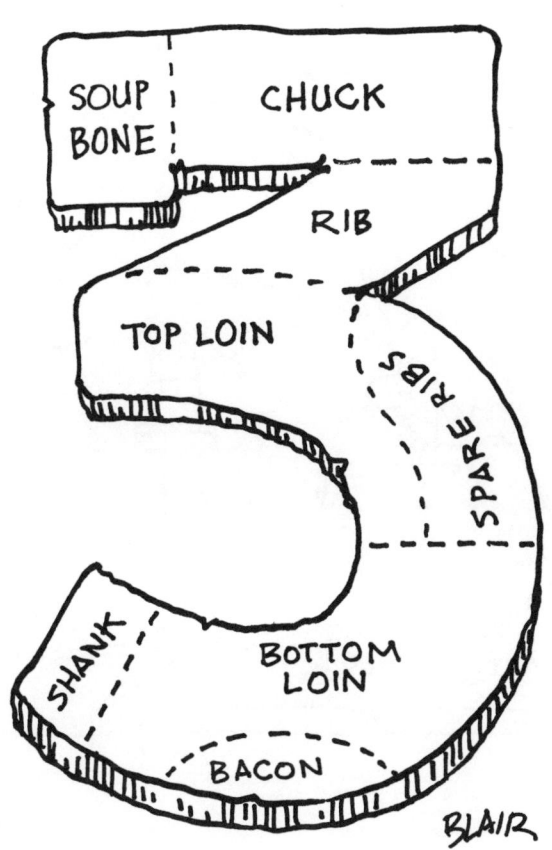

A PRIME NUMBER

" BAD NEWS. MRS. PHAROAH HAS DECIDED SHE WANTS THE PYRAMID OVER THERE INSTEAD."

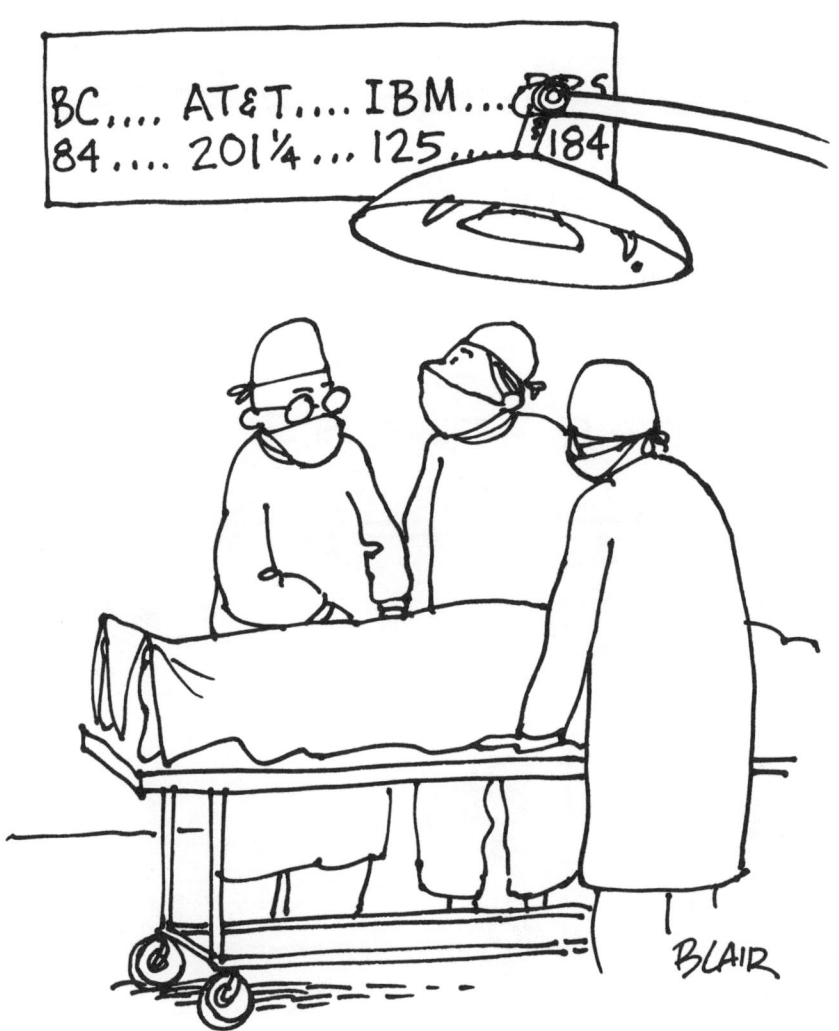

AT BAT: 24
BATTING AVERAGE: .276
SALARY: $1,750,000
THIS AT BAT: $3,275

371 FEET

"I'M NOT SURE I LIKE THIS NEW SCOREBOARD."

"... AND AFTER ONLY 10,000 MILES, YOU QUALIFY FOR AN AROUND-THE-WORLD FLIGHT."

" CUT MY PIZZA INTO FOUR PIECES...
NO WAY I COULD EAT EIGHT. "

"AS THE COLD FRONT MOVES IN, TEMPERATURES
WILL PLUNGE TO 1450°C ... BE SURE TO TAKE IN
YOUR MOLTEN METALS."

"I THINK IT SAYS ' THE SQUAW ON THE HIPPOPOTAMUS EQUALS THE SUM OF THE SQUAWS ON THE OTHER TWO HIDES...'"

COCOON

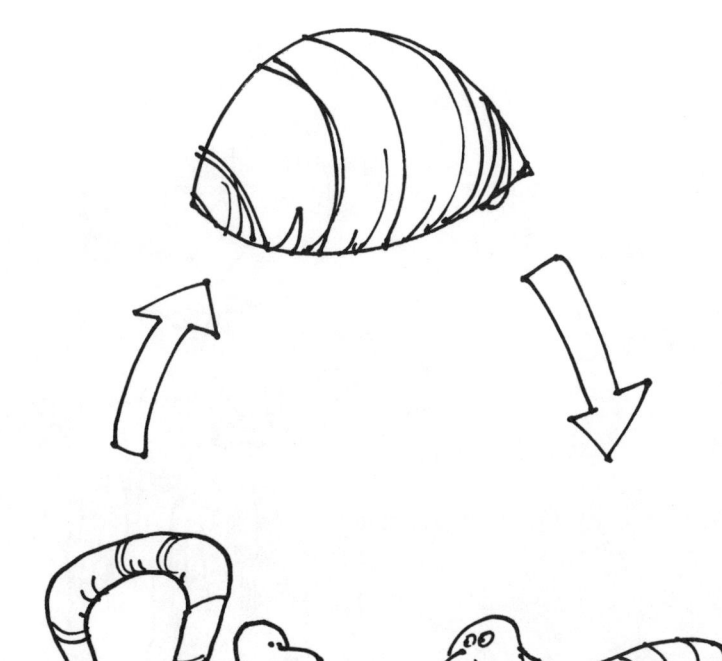

INCHWORM CENTIPEDE

BLAIR

METRICATION

"DR. ZIMMERMAN IS AN EXPERT ON CHAOS THEORY. HE'D LIKE TO OBSERVE YOUR SEVENTH PERIOD CLASS."

NUMEROLOGISTS ON SAFARI

" LET'S SEE... A TRIANGLE BY ANOTHER MONIKER STILL
HATH THREE VERTICES... A CIRCLE BY ANY OTHER LABEL
IS JUST AS ROUND... UM... A ROSE... "

"THE ONLY GRANT AVAILABLE WAS FROM THE NATIONAL ENDOWMENT FOR THE ARTS."

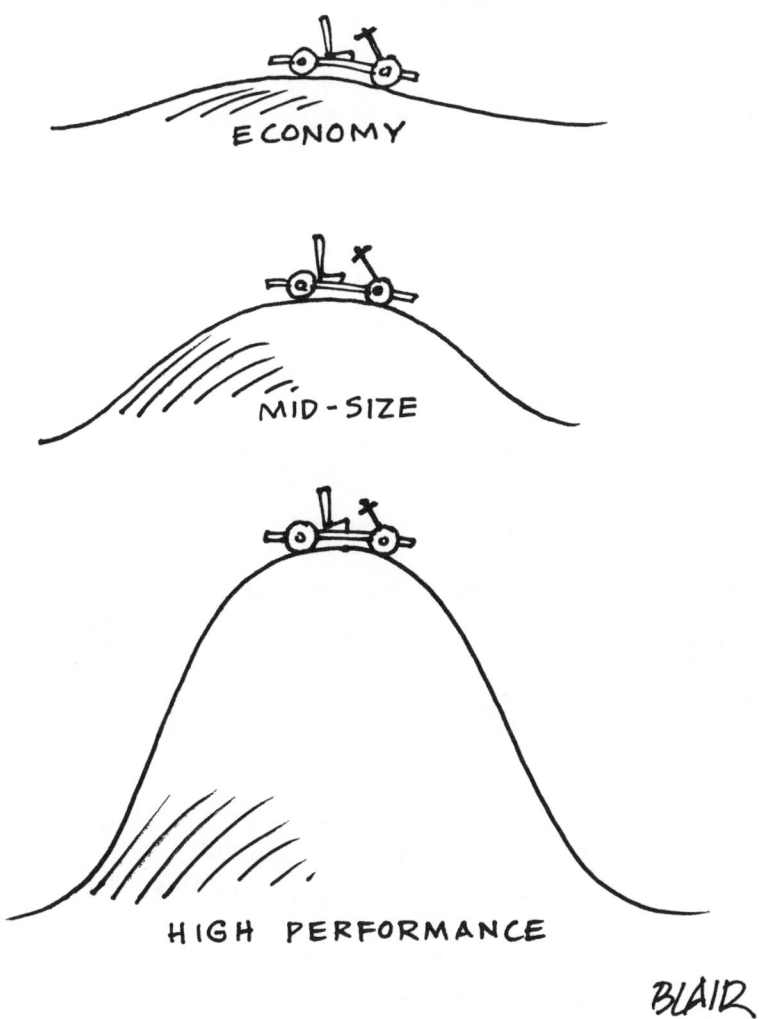

ECONOMY

MID-SIZE

HIGH PERFORMANCE

BLAIR

STONE AGE CARS

" I ASSURE YOU, MADAM, PRESTIGE ACADEMY ACCEPTS
NO STUDENTS BELOW THE FIRST PERCENTILE. "

"IT'S THE ARCHITECT'S FIRST SKYSCRAPER. SOMEONE
TOLD HIM IT'S TRADITIONAL TO LEAVE OUT THE
THIRTEENTH FLOOR."

"HERE IT IS, LEWIS! THE CONTINENTAL DIVIDE!"

WHEN MATHEMATICIANS CAN'T SLEEP

"I CAN'T PLAY THIS HOLE; I DON'T HAVE
ANYTHING HIGHER THAN A NINE IRON."

" HONEY, I PICKED UP A NEW CALENDAR AT THE
DRY CLEANERS TODAY ... WHERE DO YOU WANT IT ? "

"WE BELIEVE THE LACK OF ANY SIGNIFICANT ACCOMPLISHMENT BY THIS SOCIETY IS SOMEHOW CONNECTED TO THEIR RELIGIOUS BELIEFS."

"YOUR ANSWERS ARE CORRECT, BUT UNFORTUNATELY
YOU DIDN'T SHOW YOUR WORK."

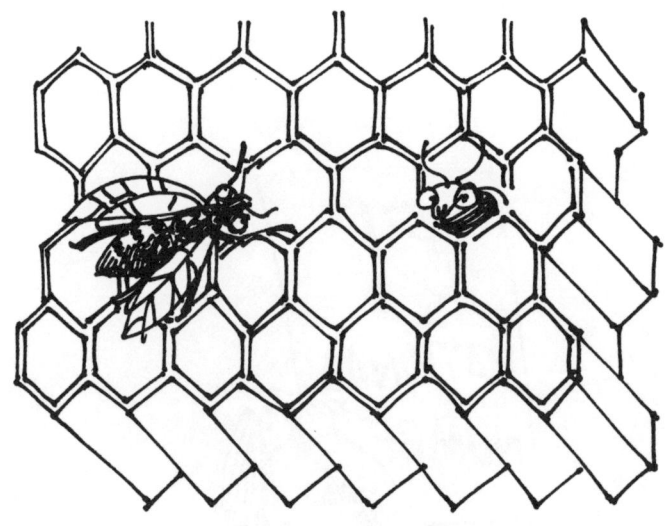

REGULAR BEES

BEE NERDS

"YES, IT'S CIRCULAR REASONING, BUT IT'S STATE-OF-THE-ART CIRCULAR REASONING."

"YOU'LL PLAY DOLLS WITH ME IF
I GUESS YOUR NUMBER, RIGHT?"

IF LINCOLN HAD BEEN AN ALGEBRA TEACHER
INSTEAD OF A LAWYER

" IT'S MY SCIENCE FAIR PROJECT,
MY PHYSICS TEACHER SAW IT AND GAVE ME AN 'A!
MY SHOP TEACHER SAW IT AND FAINTED."

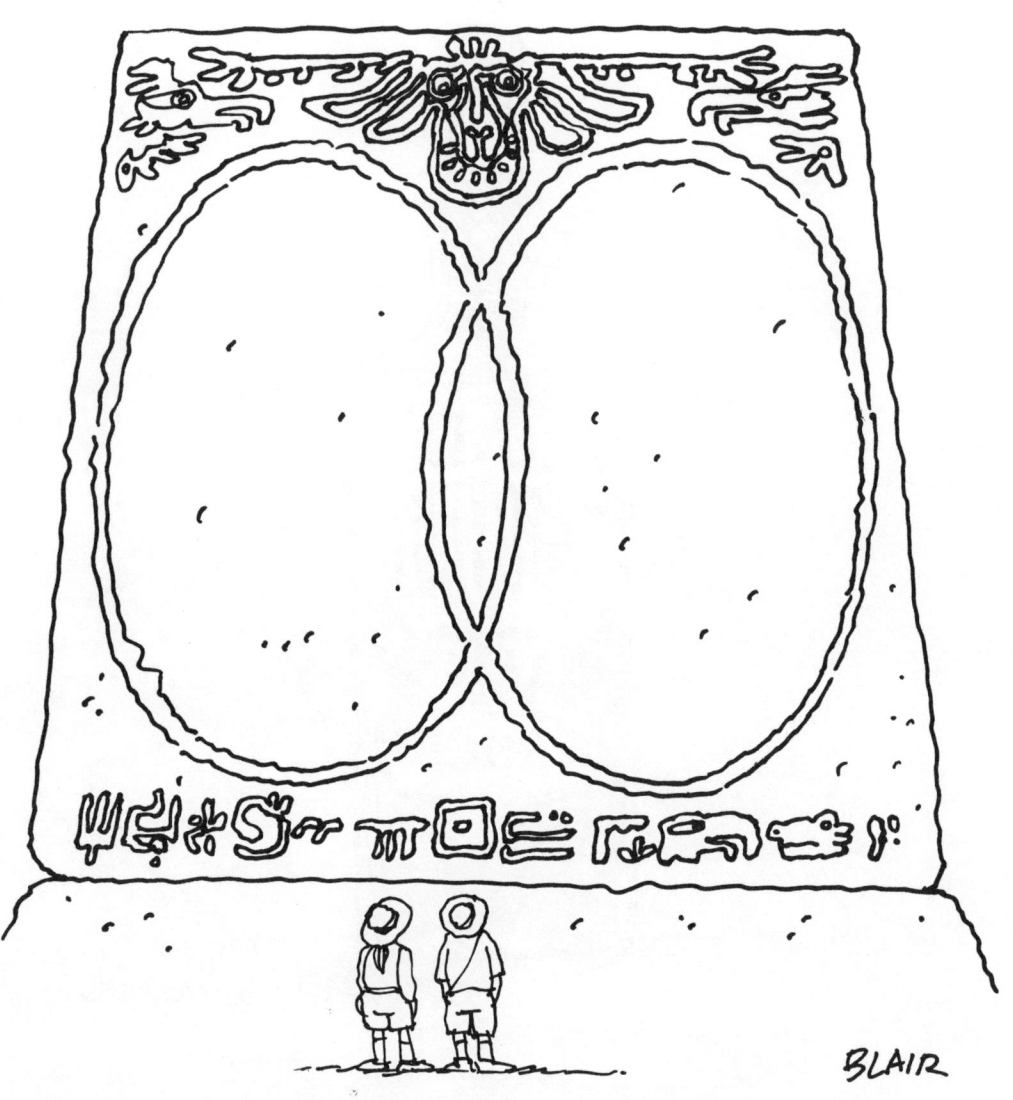

"NEAR AS WE CAN TELL, IT'S A MAYAN CREDIT CARD."

BLAIR

"AWRIGHT, BREAK IT UP. WE GOT A REPORT YOU KIDS WERE EXTRAPOLATING IN THERE."

"YOUR 2×4s ARE HERE, BOSS."

"THERE WERE THIRTEEN, BUT I MET MY EDITOR ON THE WAY DOWN."

THE INVENTION OF THE DECIMAL POINT

"BUT I ALWAYS CARRY A BOMB WHEN I FLY;
THE ODDS OF TWO BOMBS ON THE SAME PLANE
ARE PRETTY SLIM!"

"LEM, THIS TOWN AIN'T BIG ENOUGH
FOR TWO STATISTICIANS."

"IT'S SOMETHING LESTER SAW IN THE
LAWN AND GARDEN SECTION OF
SCIENTIFIC AMERICAN."

NATURAL NUMBERS,
UNLIMITED
A DIVISION OF
REAL NUMBERS, INC.

MOTHER NATURE

OUR FOUNDER

BLAIR

" YOU THINK YOU <u>MIGHT</u> HAVE MADE A SLIGHT
MIS<u>CAL</u>CULATION ??! "

MATH PHOBE IN HELL

DIRECTOR OF PROBABILITY RESEARCH

OFFICE HOURS	ODDS THAT I'M IN
8 am - Noon	5 To 2
Noon - 2 pm	25 To 1
2pm - 5 pm	11 To 2
5 pm - 6 pm	100 To 1

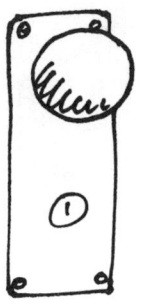

BLAIR

" THE LAB BOYS CAN TELL US FOR SURE, BUT IT LOOKS LIKE THE MURDER WEAPON WAS A .45! "

" BUT YOU SAID, 'COME READY TO DO CONSTRUCTIONS'."

" NOWADAYS YOU HAVE TO KNOW A COED'S I.Q. , HER
S.A.T. SCORE, AND HER GRADE POINT AVERAGE.
WHATEVER HAPPENED TO 36-24-36 ? "

WATER

WATER !

BLAIR

VULTURE'S - EYE VIEW OF A TRAGIC
CONSEQUENCE OF THE PARALLEL POSTULATE

BETSY'S LACK OF FACILITY WITH THE MULTIPLICATION
TABLE LEADS TO THE FAILURE OF HER FIRST DESIGN.

"FIRST HE STARTED CALLING HIMSELF 'BUCKMINSTER OF THE NORTH,' NOW THIS!"

"OF COURSE I KNOW THE DIAMETER IS TWICE THE RADIUS. WHY?"